FOOD CHAINS

Peter Riley

W
FRANKLIN WATTS
NEW YORK • LONDON • SYDNEY

First published in 1998 by
Franklin Watts
96 Leonard Street, London
EC2A 4RH

Franklin Watts Australia
14 Mars Road
Lane Cove
NSW 2066

Series editor: Sarah Snashall
Editor: Janet De Saulles
Art director: Robert Walster
Designer: Mo Choy
Picture research: Sue Mennell

Photography: Steve Shott
(unless otherwise credited)
Artwork: Peter Bull

A CIP catalogue record for this
book is available from the
British Library.

ISBN 0 7496 2966 5

Dewey classification 574.5

Printed in Belgium

Picture credits:
cover shot and title page: Planet Earth
Pictures/P.N. Raven
Bruce Coleman pp. 4l (C. Varndell), 4b (H.
Reinhard), 5 (T. Buchholz), 7m (K. Taylor),
7b (H. Reinhard), 10b (S. J. Krasemann),
11 (H. Reinhard), 13b (Jeff Foott
Productions), 15t (J. Burton),
16b (E. and P. Bauer), 18t (Dr F. Sauer),
22t (C. Fredriksson), 27b (W. S. Paton)
Robert Harding pp. 8b, 12b,
28 (D. Wilson), 29b
FLPA pp. 10t (M. Newman), 17 (D.
Hosking), 21b (R. Wilmshurst),
26l (L. West)
Images Colour Library pp. 9t
Oxford Scientific Films pp. 6
(Norbert Rosling), 7tr (N. Benvie),
9b (A. Root/Survival), 12t (P. Parks),
13t (T. McHugh), 15b (T. Andrewartha),
20l (J. Dermid), 21t (S. Dalton)
Planet Earth Pictures pp.
18b (P. De Oliveira), 22b (A. Bärtschi)
NHPA pp. 20b (S. Dalton),
25b (P. German)
Still Pictures pp. 8t (R. Seitre),
14t (M. Kazmers), 14b (Norbert Wu),
23 (A. Bärtschi), 24 (R. Seitre),
25t (F. Lemmans)

CONTENTS

EATING TO LIVE

There are millions of different kinds of living things. We divide them into two main groups – plants and animals. The difference between the two groups is that plants make their own food while animals need to eat in order to survive.

WHAT IS A FOOD CHAIN?

The way plants and animals are linked together by feeding is called a food chain. Most food chains start with energy from the Sun. Plants use this energy to grow. In turn, the plants may be eaten by animals, and these animals by other animals.

TRAPPING ENERGY

A green plant makes most of its food in its leaves. The cells inside the leaf have a green pigment called chlorophyll. This traps some of the energy of sunlight and uses it to make food.

Animals cannot make their own food. Some animals, such as puffins, feed on other animals.

The light falling on this tree shines through to the chlorophyll inside the leaves.

MAKING FOOD

Plants use their roots to take up water and minerals from the soil, and use their leaves to take in carbon dioxide from the air. Inside the leaf, energy from sunlight is used to change the carbon dioxide and water into sugar and starch. This food-making process is called photosynthesis. Food is stored in the roots, stems, leaves, fruits and seeds. These plant parts are the food of many animals.

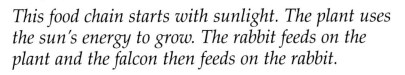

This food chain starts with sunlight. The plant uses the sun's energy to grow. The rabbit feeds on the plant and the falcon then feeds on the rabbit.

FOOD CHAINS AND FOOD WEBS

A food web, such as the one shown below, shows many food chains woven together. It illustrates how animals in one environment feed off the animals and the plants available to them.

Carrots store most of their food supply in their roots.

HERBIVORES

Animals have adapted to eat the food that they find around them. Some animals survive by eating only the plants in their environment. Such animals are called herbivores.

GRAZERS AND BROWSERS

Few herbivores eat all of a plant – each animal usually eats just certain parts of a plant. Some animals eat only the leaves of plants. Large animals that eat leaves can be divided into grazers and browsers. Grazers such as the zebra eat grass, while browsers such as the giraffe eat leaves on trees. Both kinds of animal have large back teeth for grinding up their food.

With their long necks, giraffes are able to browse on the leaves high up in trees.

SEEDS AS FOOD

Many birds eat seeds. A seed contains a tiny plant and its store of food, and is protected by a tough seed coat. Birds such as finches have strong beaks to break the seed coat and reach the food.

Seeds are protected by a hard coating.

PLANT JUICES AS FOOD

Some animals feed on plant juices. Bees and butterflies land on flowers to feed on their nectar. This sweet liquid gives the insects energy to live and to fly. Greenflies have special tube-like mouths with which they suck the juices out of plants.

This butterfly is feeding on flower nectar.

STAYING SAFE!

Because herbivores are the food of other animals, they have developed ways of avoiding being caught. Many herbivores, such as the caterpillar, are camouflaged so it is difficult to see them. Some herbivores, such as the finch or the rabbit, move quickly to try and escape if they are discovered.

▌INVESTIGATE!

Look for caterpillars and aphids on plants. What part of the plant do you find them on? Do not touch the caterpillars – some are poisonous!

These caterpillars are a similar colour to the leaves they are feeding on. This makes it difficult for predators to see them.

The next link in the food chain is between herbivores and the animals that eat them. Such animals are called carnivores, or predators, and the animals they eat are called their prey.

Anteaters use their long, pointed noses and sticky tongues to feed on ants and termites.

RANGES OF FOOD

Some carnivores can only feed on a few kinds of food. The anteater cannot open its jaws but it can shoot out a long sticky tongue to catch ants and termites. Other carnivores eat a wide range of food. The jaguar, which lives in the South American rainforest, eats fish, turtles, alligators, deer, anteaters and monkeys.

CARNIVORES' TEETH

The teeth of a carnivore are shaped to deal with their food. Lions and wolves have large fang teeth, called canines, for stabbing their prey. Between the canines are incisors, which are used to cut into flesh. At the back of the mouth are carnassial teeth for cracking open bones.

A lion displaying its sharp incisors and canines.

_This eagle is poised
to drop on its prey,
capturing it with its
large talons._

CARNIVOROUS BIRDS

Carnivorous birds, such as the
eagle, have large claws on their
toes, called talons, for stabbing
and gripping their prey. They
also have a hooked beak for
tearing up the flesh of their prey.

POISON!

Carnivores such as centipedes, spiders and
some snakes inject poison into their prey
when they bite. The poison stops the prey
from struggling and kills it quickly. The
predator is not affected by eating its
poisoned prey.

_A snake feasts on a
smaller snake!_

LONG FOOD CHAINS

There may be several links in a food chain: a
jaguar eating an alligator is an example. Before
the alligator was caught, it may have eaten a rat,
and the rat may have eaten a spider before it
was caught. The spider, in turn, may have eaten a
fly, and the fly may have fed on a plant.

▌INVESTIGATE!

Watch a dog or a
cat eat. How does
it use its teeth?

OMNIVORES

Most animals eat either animals or plants. Animals known as omnivores, however, eat both plants and animals. Such animals are well adapted to their environment and rarely go hungry.

This brown bear is holding its prey between its teeth.

TEETH AND CLAWS

Omnivores' teeth are less specialised than those of herbivores or carnivores because of the many types of food they eat. Brown bears live in forests in North America, Europe and Asia, and use their teeth for feeding on moss, berries, fish and deer. They use their claws to tear open a wild bees' nest for a meal of honey or grubs, or to help kill deer.

FEEDING EVERYWHERE

Omnivores may search in many places for their food. The North American raccoon lives in woods and thickets near water. It digs up worms and insects from the soil; paddles in ponds and rivers to find fish, frogs and crayfish; climbs trees to find eggs and chicks – and feeds on any fruits and berries it finds. Some raccoons have adapted to living in towns, scavenging for food in bins.

A raccoon searches for food near the water's edge.

EATING WHAT THEY FIND

The wild boar lives in forests in Europe. It feeds in just one part of the habitat – the ground – but it will eat almost everything it finds there. It uses its snout to dig in the soil and find roots, acorns and insects. As it searches the forest floor, it may catch mice, voles and lizards, or eat the eggs of ground-nesting birds.

The wild boar eats an amazing variety of food, including roots, acorns, beetles, mice, eggs, lizards and voles.

Crayfish are found in many rivers all over the world.

TAKING ALMOST EVERYTHING

Some omnivores, such as the crayfish, not only eat living plants and animals, but they feed on their dead bodies too. Crayfish eat water plants, worms, snails and tadpoles, but will also eat dead plants and the bodies of dead animals. Crayfish are cannibals – large crayfish will eat smaller ones.

INVESTIGATE!

Are you an omnivore? Make a list of the foods you eat. Which ones come from plants and which come from animals?

IN THE OPEN SEA

The sea is home to many different plants and animals. As on land, food chains start with the action of sunlight on plants. The tiny plants in the sea are called phytoplankton. They float near the surface of the water so the sunlight can reach them.

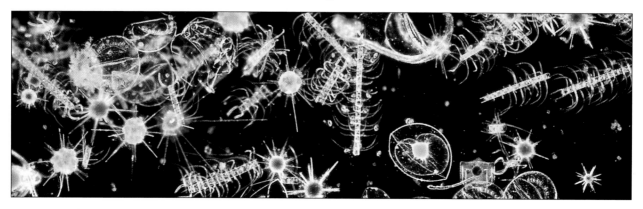

A view through a microscope of phytoplankton and zooplankton.

ZOOPLANKTON

Small animals, known as zooplankton, float or swim with the phytoplankton. Some of the tiny animals are called copepods. They feed on the phytoplankton. Copepods have hair-covered limbs which they wave in the water to make water currents. These carry the phytoplankton to the copepods which then catch and eat them.

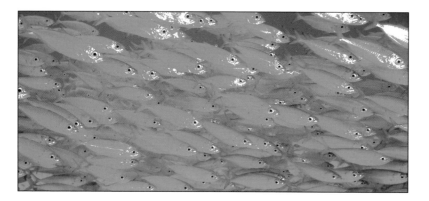

FISH

The smaller animals in the zooplankton are eaten by larger ones, and the larger ones are eaten by fish. Shoals of fish, such as herring, swim through the plankton to feed. Each fish opens its mouth and lets water pass through spines on its gills. The plankton are trapped on these spines and are swallowed by the fish.

A shoal of herring.

LARGE FISH

Large fish, such as tuna, feed on smaller fish. They swim into a shoal, catch the fish in their jaws and swallow them. Sharks either swallow their prey whole or tear it into chunks.

A shark swimming through the water, carrying its prey in its mouth.

Humpback whales use the strips of bristles in their mouths, or whalebone, to filter krill from the seawater.

KRILL

Plankton are also eaten by a small, shrimp-like animal called krill. Huge shoals of these animals form and are eaten by seabirds, seals and whalebone whales such as humpbacked whales. A whalebone whale may be attacked and, in turn, eaten by a pack of up to forty killer whales.

■ INVESTIGATE!

Watch any kind of fish being fed in a fish tank. How do they take in the food?

13

ON THE SEASHORE

Food chains on the seashore may start in two ways. They might start when the tide comes in – with the plankton in the seawater – or they may start with seaweeds growing on the rocks.

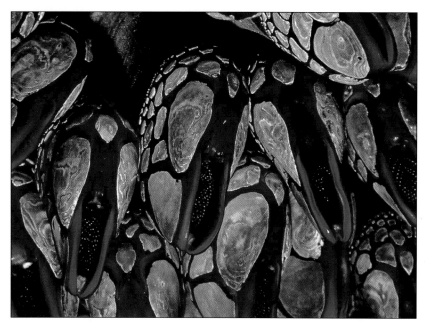

These goose barnacles feed by opening their shells so they can filter the water for food.

CHANGING TIDES

When the tide is out, barnacles and mussels attached to rocks keep their shells shut to keep themselves from drying out. When the tide comes in, they open their shells a little and filter the seawater for plankton. Starfish, in turn, feed on mussels. They use their arms to pull the mussel shells apart and then make their attack.

SEAWEED

Seaweed grips the rocks with a root, or holdfast. This stops it being washed away. Seaweed belongs to a group of plants called algae. In some places, many tiny algae plants grow close together over the rocks, covering them like a green skin.

Sea snails feed on the seaweeds, using their tongues which are covered in teeth. They rub their rough tongues over the algae to break up the plants and feed on the pieces.

A sea snail feeding on seaweed.

IN THE WATER

The sea anemone has tentacles covered in stinging cells. When a shrimp or a small fish touches them, the stinging cells fire a poison into the animal's body and kill it. The sea anemone pulls the body into its mouth with its tentacles.

This fish is about to be eaten by an anemone.

Oystercatchers feed on cockles, mussels, limpets and crabs.

ON THE SHORE

Shore crabs eat anything they can catch when the tide is in. They hide away at low tide – but hungry gulls still manage to find and eat some of them. The gulls hunt along the shore, pecking at the crabs and hammering open shellfish. They also search rock pools where they catch fish in their long, sharp bills.

■ INVESTIGATE!

Algae does not only grow on the seashore. It forms a green coat wherever it is damp. Look for it growing on stones and walls in your neighbourhood.

FOOD CHAINS ON GRASSLANDS

Many food chains around the world start with grass. The grasslands of North America, Africa and Australia each have their own unique food chains.

NORTH AMERICA

The meadows of North America and Europe are home to field voles. These animals look similar to mice and eat the lower part of grass stems. Field voles make tunnels through the grass to hide from hawks and owls, their predators.

On the plains, prairie dogs make burrows in which to hide from predators such as coyotes and hawks. They eat the grass around their home so they can see any approaching attackers.

Prairie dogs feed on grass.

THE AFRICAN PLAINS

Large numbers of zebras, gnus and Thomson's gazelles graze on the African plains. Zebras eat the tops of the grass first, then gnus eat the middle part of the grass. Thomson's gazelles eat the young shoots which are left behind. All these herbivores fall prey to lions, hyenas and other carnivores.

THE AUSTRALIAN OUTBACK

In Australia, grasses are eaten by kangaroos and wallabies. These herbivores may be eaten by packs of wild dogs called dingoes. Grass seeds are eaten by diamond firetail finches, which in turn are eaten by the peregrine falcon. The falcon dives out of the sky to attack and feed on the finches.

Kangaroos feed on grasses and leaves.

In this scene, auger buzzards fly high above the African plains searching for prey, while a hornbill and some weaver birds sit in the branches of an acacia tree. Giraffes, zebras and elephants roam the land looking for plants, and a lion gets ready to attack a klip-springer.

FRESHWATER FOOD CHAINS

Fresh water forms streams, rivers and ponds, and is not salty like seawater. Each kind of freshwater habitat has its own food chains.

STREAMS

In streams, caddis-fly larvae build nets under the stones to trap any small animals and plants that are carried along by the fast-moving water. Small fish, such as the European bullhead, feed on tiny animals under stones – including the caddis-fly larvae. The fish may be eaten by the kingfisher, which dives into the water and spears them with its sharp, pointed beak.

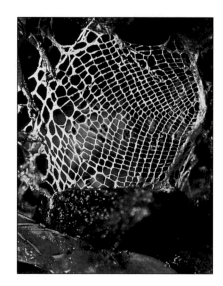

The caddis-fly covers its body in stones and makes a net to catch food.

A kingfisher dives for food.

RIVERS

The water moves slowly in some parts of a river. This gives water plants a chance to take root and grow. Snails eat water plants and fish eat the snails. A large bird called a heron eats the fish. It stands so still that fish do not notice it and swim close by. When one of the fish is very close, the heron plunges its head into the water and catches it in its long, sharp beak.

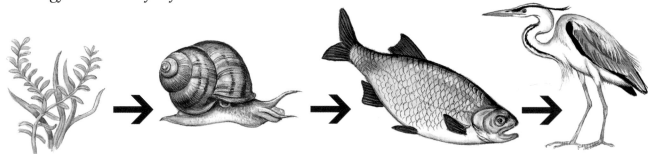

In a slow-moving river, herons eat fish, which eat snails. These, in turn, eat weeds.

PONDS

On the still surface of a pond, whirligig beetles spin round, eating the insects that fall into the water. Fish and water birds eat the beetles.

Shortly after tadpoles have hatched from frog-spawn, they start to feed on the algae which covers the surface of water plants. Later, they eat small water animals. Fish and water birds eat so many of the tadpoles that only a few survive to grow into frogs.

■ INVESTIGATE!

Make a pond in a jar with water plants, snails and water fleas called daphnia bought from a pet-shop. Compare how the water fleas and snails move.

A pond with water lilies, reeds and sedges. A duck swims on the surface and a heron searches for fish. Loaches lie at the bottom, while a carp swims higher up in the water.

WOODLAND FOOD CHAINS

There are woodlands in many parts of the world. The trees store food in the wood they make, and carry water from their roots to their leaves in tiny pipes in their trunks and branches.

TREE RINGS

The pipes make rings in the wood every year. When the weather is good, the tree makes a large amount of food and grows a wide ring. When the weather is poor, the tree makes a narrow ring.

WOODLAND INSECTS

Bark beetles burrow underneath the bark, feeding on the rings of wood. Woodpeckers prey on bark beetles, striking the wood with their chisel-shaped beaks. They use their long tongues to lick the beetles out of their burrows.

Some spiders spin webs between twigs. They hide on a twig, waiting for an insect to fly into the web and become stuck on its threads. As the insect struggles, the spider rushes up to it and gives it a poisonous bite.

The spider crawls on top of its captured prey.

BATS

Moth caterpillars feed on leaves. They are camouflaged, so many of them escape predators. When they become moths, they fly at night, trying to avoid bats. These predators roost in tree holes during the day and fly through the woodland at night, feeding on moths and other insects such as gnats and beetles.

This bat is about to scoop up a moth for its meal.

WOODLAND BIRDS

The tree creeper bird feeds on insects and spiders, using its hooked beak to pull them out of their hiding places.

▮ INVESTIGATE!

Look at the rings on the end of a cut branch or a tree stump. Were there any years when the tree grew extra well?

The tree creeper's camouflage keeps it safe as it searches for food.

Rainforest Food Chains

Rainforests are found in Central and South America, South East Asia and in West and Central Africa. They contain an amazing variety of food chains.

The Colobus monkey lives in the canopy of the African rainforest.

The emergent layer
The top layer of the rainforest is made by the tallest trees and is called the emergent layer. Birds of prey nest here.

The canopy
The canopy lies below the emergent layer and is made by the leaves and branches of slightly smaller trees. Fruit from the trees is eaten by birds such as the hornbill of Malaysia, and by animals such as monkeys and squirrels. In the South American rainforests, humming-birds feed on the nectar produced by flowers of the canopy.

The understorey
The understorey is formed from even smaller trees. Huge numbers of insects eat the leaves and fruits of the plants at all levels in the rainforests. Tree frogs live in the understorey of the South American rainforest and feed on the insects living there. Predatory birds and snakes eat the frogs.

The poison arrow frog has poison in its skin and warns predators not to attack by flashing its bright colours.

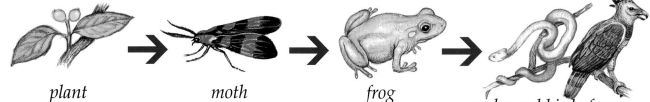

plant   moth   frog   snake and bird of prey

Parrots feed on fruits, seeds and nuts. Their droppings fall to the forest floor, spreading undigested seeds and allowing new plants to grow.

THE FOREST FLOOR

The leaves and branches of the trees stop most of the light reaching the ground. Few plants grow here and the ground is covered with fallen leaves and fruits. These are eaten by worms, snails and millipedes.

RAINFOREST RIVERS

Insects which fall into the water are eaten by many different kinds of fish. In South America, large animals entering the water may be attacked by piranha fish or alligators. In other rainforests, crocodiles may pull animals in from the bank and drown them.

INVESTIGATE!

Plants grow fast in the warmth of a rainforest. Try growing two sets of pea seedlings, one in a warm part of a room or garden, the other in a cold part. Do they grow better in the warmth too?

DESERT FOOD CHAINS

Water is essential for life. In a desert it may rain for only a few days a year, so the desert plants have adapted to survive without much water.

SURVIVING WITHOUT RAIN

Since plants lose water through their leaves, desert plants have only small leaves or have spines. Some plants survive the dry conditions by remaining as seeds while it is dry. When it rains, they germinate and grow quickly into plants. Some plants have long roots to collect as much water from the ground as possible.

Camels rely on plants as a source of water.

Desert animals are also able to survive without much water. One of the best-known desert animals, the camel, lives in the Sahara. It can manage for weeks without drinking any water, as long as it can find succulent desert plants to eat. These are plants which store water in their leaves and stems.

DESERT PLANTS AND FOOD CHAINS

Because of their stores of water, desert plants provide the basis for many desert food chains. Insects which eat desert plants include locusts, grasshoppers and crickets. These animals, in turn, provide food for lizards, birds and scorpions.

In this North American desert scene, a rattle snake watches a tortoise while a kit fox closes in on a horned lizard. A cactus wren and a scotts oriole sit on some cacti and a burrowing owl and cottontail rabbit rest on the ground.

Huge swarms of locusts feed on leaves – they can strip plants bare within seconds.

The insects in Australian deserts are eaten by many different kinds of lizard-like reptiles, such as this gecko.

■ INVESTIGATE!

Germinate three groups of cress seeds. Every day, give the first group a large amount of water, the second group a medium amount, and the third group only a little water. Which group grows best?

25

RECYCLING FOOD MATERIALS

Nature constantly recycles its materials. When an animal or plant dies, it rots, or decomposes. In the same way, the waste animals produce from undigested food are broken down into simple substances. These substances are then used by living things.

As these leaves rot, they will start to become part of the soil.

These toadstools were formed by fungi feeding on rotting matter.

FUNGI

A fungus is neither a plant nor an animal, but is a third kind of living thing. It feeds on the first two groups by growing tiny threads into their dead remains, causing them to dissolve. This plays an important part in how dead things decompose.

The dissolved mixture is drawn up the threads and lets the fungus grow. When the fungus is fully grown, it may produce a toadstool. This makes tiny spores which float away on air currents. When a spore lands on more remains, a thread grows out of it and it starts to feed.

BACTERIA

Very small living things, called bacteria, also feed on plant and animal remains. Along with fungi, they help break down the remains into simple substances.

As they do their work, the bacteria release carbon dioxide into the air. This is taken in by plant leaves and used in photosynthesis. Minerals are also released from the remains by the action of the bacteria. These minerals dissolve in rain water and are taken up by plant roots to make more food.

Even a small amount of soil contains millions of bacteria.

BUILDING NEW LIFE

All living things are broken down into basic minerals or chemicals when they die. These provide the building blocks for new plant and animal life. In this way, all life around us is actually made up of recycled materials!

Both this rabbit and the grass it is feeding on are made of recycled materials.

YOU AND FOOD CHAINS

You probably found out from the Investigate! section on page 11 that you eat food from lots of different plants and animals. This means that you are part of many food chains.

BREAD

Most bread is made from wheat. Like other plants, wheat takes in sunlight and carbon dioxide from the air, and water from the soil.

We eat the grains of the wheat plant. They contain the wheat seeds and are ground up to make flour, which is then used to make bread. This is one of the simplest food chains of all. It is wheat ➔ you.

These grains have grown so large that the wheat stalks are bending under their weight.

Eggs are laid by hens which may eat pelleted food made from a range of foods, including maize.

EGGS

A slightly longer food chain is made when you eat an egg. The hen that laid the egg probably ate the grains of a maize plant. The food chain here is maize ➔ hen ➔ you.

MILK

Another two-link food chain is made when you drink milk. The cow that made the milk ate grass. The food chain is grass ➔ cow ➔ you.

FISH

If you eat fish, the food chain is much longer. It is phytoplankton ➔ zooplankton ➔ small fish ➔ large fish ➔ you.

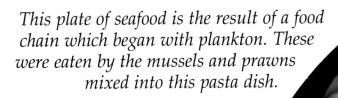

This plate of seafood is the result of a food chain which began with plankton. These were eaten by the mussels and prawns mixed into this pasta dish.

■ INVESTIGATE!

Work out food chains for ten foods that you eat. How long are most of the food chains?

GLOSSARY

ADAPT – when an animal or plant changes in order to survive in a particular country or environment. Racoons, for example, have adapted to living in towns by learning to look for their food in rubbish bins.

APHID – a bug-type insect such as a greenfly or blackfly.

BACTERIA – tiny living organisms that can only be seen with a microscope. They have simple bodies which may be shaped like a rod, a ball or a spiral. Bacteria are found almost everywhere – in the air, soil and on the skin.

CAMOUFLAGE – the means by which animals hide against their environment. For example, green caterpillars blend in with the leaves they feed on, staying hidden from predators.

CARBON DIOXIDE – a gas which is released into the air by animals and plants. Animals release it when they breathe. It is also taken in by plants and used to make food.

CARNIVORE – an animal that feeds only on other animals.

CELL – a tiny part of the body. The bodies of large plants and animals are made from millions of cells. There are many different types of cells.

CHEMICALS – the basic building blocks of all life.

CHLOROPHYLL – a green substance in plants which traps some of the energy in sunlight, allowing the plant to make its food.

COPEPOD – a microscopic animal related to shrimps and crabs.

DECOMPOSE – to rot or decay. When a dead plant or animal decomposes, it breaks down into simple substances known as minerals. These minerals are then used to help new life to grow.

ENVIRONMENT – the surroundings in which a plant or animal lives. It includes both the general area and all the other plants and creatures which live there.

FOOD CHAIN – the way in which plants and animals are linked together by feeding. A food chain shows how an animal may feed on a plant, for example, but in turn become the food of another animal.

FOOD WEB – a collection of food chains which are linked together because they share one or more living things.

FUNGI – living organisms that feed on the remains of plants and animals.

GERMINATE – when a seed germinates, it comes to life, sending out shoots and growing into a plant.

GILL – a structure in a fish which allows it to breathe in oxygen from the water and breathe out carbon dioxide into the water.

HERBIVORE – an animal which feeds only on plants.

MINERALS – chemicals such as those in fertiliser. They are found in the soil and are taken in by plants to help them grow healthily.

NECTAR – a liquid containing sugar that is made by flowers.

OMNIVORE – an animal which feeds on both plants and animals.

PHOTOSYNTHESIS – the process in which green plants use energy from sunlight to make food, using oxygen from the water in the soil and carbon dioxide from the air.

PHYTOPLANKTON – the part of the plankton made up from tiny plants.

PLANKTON – living things which range in size from microscopic plants and animals, to fish eggs and small shrimps. Some plants and animals form plankton in fresh-water lakes.

PREDATOR – an animal that kills and eats other animals.

PREY – an animal which is killed and eaten by a predator.

SEED – a structure produced by a flower which contains a tiny plant and food store. In damp, warm conditions, the seed breaks open, germinates and grows into a plant.

SPORE – a very tiny case containing a small piece of a living thing such as a fungus. Spores travel in the air and if one settles in suitable conditions, the spore breaks open and the living thing inside begins to grow. Fungi reproduce by releasing spores.

SUCCULENT – succulent plants store water in their thick, spongy leaves and are commonly found in deserts.

ZOOPLANKTON – the part of the plankton that is made up of animals. The animals may be very tiny creatures covered in microscopic hairs or be small shrimp-like creatures.

INDEX